The Digital Recovery of the city of New Orleans.

Helping post-Katrina New Orleans get back online.

by

Michael William Fredrick

Introduction.

People love a journey. We love a quest, we cheer for underdogs and champion overcoming adversity.

My story isn't unique to New Orleans, but it's mine. It happened and it made me stronger, hungrier, and cemented why I'll never leave this city.

To this day, the post-Katrina website and email recovery was the largest and longest effort of my career. I'll never experience anything like that again.

It took a lot effort to get New Orleans back up and running. I had one goal in mind as I drove back into the war zone: help businesses and organizations affected by the storm. Help them reclaim their lives, even if only online. It was the only way I knew how to help. Sort of.

I had no idea what I was doing and more importantly, what I was in for. I was a web designer. I built websites and hosted them. That's it. What did I know about recovery after a hurricane?

1. So, you're a computer nerd, huh?

My family moved a lot. New schools, new people, and new towns were a scenario I was used to. But, while the locations changed, I had one constant: computers.

Eventually, we settled in Memphis. By then, I was a bonafide computer nerd. This was the infancy of the Internet with AOL, Compucast, and I was hooked. Throughout high school, I was the typical bored computer teenager, more interested in programming and design then studying and grades. It's cliche, but cliches exist for a reason.

College didn't interest me. I jumped straight into the workforce. This meant a meager salary at a call center job supporting cable modems. Not exactly Silicon Valley, but it was a job with computers. Naturally, as any good screw up, I didn't last long. Because of my ADD nature of constantly having to experiment, I needed to challenge myself. I needed something more than troubleshooting why someone wasn't getting their grainy porn fast enough.

I started experimenting with Macromedia's Flash software. Flash seemed to be where things were headed. The Internet was evolving quickly.

We could watch videos online, or we could see animated graphics on websites. Someone had to program that stuff. I figured if I wanted to get the hell outta here, I needed to learn something new.

Otherwise, I was going to get stuck talking Aunt Ester down when she couldn't get her email to work forever.

Luckily, the computers at the call center weren't exactly locked down, so I installed Flash development tools. Since I was working graves, I was able to teach myself how the coding worked. I went nocturnal, even on the weekends, hanging out on Highland Avenue at the late night coffee shop.

Immediately, I fell in love with Flash and soon began building sites.

Web design took on a structure in my head:

File libraries + Image libraries +Javascript = website.

Next, I learned HTML/CSS. Before I knew it, I had a few sites in my portfolio. People wanted me to help them get online. And they were paying me for it. My head was spinning.

And then, I moved to Los Angeles. Some friends I had met online asked me if I'd be interested in working in California on some major design and development contracts. Living in Memphis my entire life was never the plan. I gave away everything car and hit the road.

I was living in LA and dropped right in the middle of the startup scene. This was *NOT* Memphis.

There was wine and cold beer in the fridge, those fancy chairs, everything that comes to mind when you think thriving dot com company.

I learned more about design on that job than I ever did reading books and blogs. I owe everything to my time out in California because it made me a better designer and made me better at my job's essential function: to serve

4

the people. But soon, I was looking out past the palm trees and wondering what was next? Growing up in the South and moving to the West coast didn't fulfill me. Los Angeles was exhausting.

Being young and idealistic, I decided that I didn't need a boss and project managers. Oh, fickle youth what a cruel mistress you are.

In my head, I figured: I'm the designer, I'm the developer, I'm the system administrator, I'm doing this on my own, and I'm making money.

Screw Los Angeles, I'm moving to New Orleans.

It sounds insane and idealistic because it was. But, let's be real: most people aren't skipping out on a profitable career in LA for New Orleans. No, that move was based on one thing alone: love. The heart wants, what the heart wants and I wanted to be in New Orleans.

I wanted to seize my life's dream and open a Web design company in New Orleans.
I moved down, got an apartment and took any gigs I could get. About the same time, my father opened a business called point2point back in Memphis. It was a

structured cabling company, requiring the exact skill set I was equipped with, and it started getting me quality leads on people who needed a Web presence. Because of this, I was driving back to Memphis, a lot. I'd collect the sites and head home to New Orleans where I designed and developed them.

Business in the Crescent City was decidedly not great, and I was dead broke. After eight months of ramen, I gave in and moved back to Memphis. All of my business was there. I can remember that heartbreaking return drive. I felt like I failed my dream. New Orleans wasn't responding to my notion of getting everyone online.

Despite the heartbreak, moving back home was the smart move. I got better at site design, and I produced on a whole new level. I was working with bigger clients who had budgets. Using the point2point banner and office and its resources, I had a legit Web design and hosting business going. But, I still had my eye on New Orleans. For now, I was licking my wounds.

2. Katrina

I remember watching the news. Folks on their roofs, Canal Street under water, the cars jammed up during contraflow, and a lot of fear of the unknown. Everyone remembers. There are no short memories of what happened to New Orleans when Hurricane Katrina nearly wiped it out. My heart broke for the city. I wanted to be there. I wanted to be with all of the friends I'd made. I wanted to do whatever I could.

The television was a non-stop horror show of suffering, and here I was, doing absolutely nothing to help.

About a month later, my phone rang. It was a friend from New Orleans, a guy named Bruce. Bruce asked if I still hosted websites and if I'd be interested in helping a few customers of a business he used to work for. The company lost everything in the storm. There were a few high priority clients who needed their sites back online ASAP. I told Bruce to get me what he could. I'd do my best.

Amazingly enough, as many Katrina stories go, Bruce waded into the server room up to his chest in water and recovered the hard drives with the site files and databases. He boxed everything up and shipped what he had to Memphis.

When I looked into the hard drive, I saw 300+ folders of Web sites stored there. Getting the files was half of the battle.

The previous company had registered all of the domains with his business's name and contact information. These were locked away in his account, which meant that all of the domain names were frozen in their current status.

The problem was, there was no way to log into the company's site to make changes to the DNS or update contact information... because their servers were underwater.

The business owner gave me the registrar login and the hosting files as long as I gave him a few of the domain names in his account and sent him back data on the hard drive Bruce mailed me. We had several hundred domains, all registered to a hosting provider who was no longer in business. There was no financial data since that was in the same building with the soaked servers. This job was like solving a Rubix cube - a new challenge presented itself daily. How were we going to break into the backend of these sites?

This was going to require some sleuthing:
Every site has a contact page. Every contact page has an email address. Oh, that's right, the email server got soaked too, so we're going to have to call. Scratch that. What about phone numbers on contact pages?
Those are always there. Yeah, let's start there. It was going to be tedious, but if it worked, I had somewhere to begin, I'd have a map to follow.

When viewing a bunch of directories, the numerical directories show up at the top. I dragged over the folder, located the contact.html file and dialed the number. Someone answered. I couldn't believe it. It worked.
I explained I was attempting to get their Web site and email back up and running.

It'd been over a month since the storm, and the client was ecstatic. I copied over their files, made some minor adjustments on the contact form page so that the email form worked, and set up a few email accounts. I did it. I'd gotten someone in New Orleans, post-Katrina backup and running. I copied over the next site and dialed the next number. They were ecstatic too. This felt good.

The real fun though, was that I had access to the DNS of domain names and servers. I was often able to turn a site back on while speaking to the customer. This was due to the super fast DNS refresh of the registrar but proved to be priceless in getting folks back up.

3. Rome- er, New Orleans wasn't built overnight

After doing some math and realizing the time it would take to call over 300 businesses and organizations, I needed help. I tossed up an ad on craigslist describing exactly what we were doing. I needed someone patient and someone ready to make a lot of phone calls. Peter M., the first official employee of the point2point empire joined my ranks. I paid for a Vonage landline with a 504 phone number to make the recovery process easier, and we started calling.

Eight hours a day, five days a week, we left voicemails. We got cussed at because people thought we were a scam. Hangups, more cursing, and more voicemails, that was our daily routine.
It was a coin toss every time we called someone. They were either thankful and happy to hear from us or angry

because they thought we were the ones who abandoned them.

Early on, I decided I would honor the former host's pricing for a domain renewal and hosting. It wasn't worth haggling over. Not with people who'd lost so much. Most of the Websites I recovered were static HTML sites with very little technology in the backend. Now the sites that were not covered in the "most" area were incredibly complex PHP/MySQL websites that were custom build for a server - a server that currently had mold growing in it. (Gross, right?) But, despite this being way out of my league of expertise, I went ahead tried to save these sites, anyhow.

Let's be honest about this: My background was primarily Web design with just the basics of what's involved with migrating custom work from one server to another. By NO means should have been allowed to mess around with these sites.

I look back on it and realize I should have managed expectations better. Navigating through custom PHP code and trying to cram a database into a server can be a complete act of futility, but you don't realize that until you realize it. I'm proud of what I did, but you can't save every patient. Luckily, many of the sites were static sites that can be hosted on a potato attached to a network cable.

4. Unforeseen Issues

Believe me. This whole recovery process wasn't a cakewalk. We weren't killing it left and right. As we know with many things in New Orleans, the best-laid plans don't always work out. Despite getting sites back up and running, we still had to convince people we were legit. I was still stuck in Memphis, and I still wanted to move down to New Orleans. Yes, even after the whole destroyed city thing.

Roadblocks became the norm after we closed out about little more than half of the sites. More than once there were issues that I couldn't work my way around and just had to plow through.

One of these problems was dealing with the aftermath of a company based out of New York. This was a major headache. The company was also doing site recovery but quickly figured out that the legwork we were putting in wasn't worth it. So, they split, and we picked up the slack. It sounds easy enough, but really, it was a mess.

We had customers who'd now spoken to two different individuals in two different states and now had to decide which one to believe. Pleading our case had become well scripted because of thanks to this other company, we had to go through hoops to prove we weren't conmen.

We returned the customer's calls, and the other company didn't. We emailed back. We took action. This other company quit at the first sign of trouble AKA dealing with upset Louisiana folks.

The other painful thing was domain names expiring without any contact information. I had no idea if these people were recovering or not. Phone calls went unanswered, so we had to roll the dice. We had to pick and choose which expired domain names we'd support or let go. In those days we asked ourselves some tough questions like, Is this a domain name that should be renewed because it's worth something to someone? I often wonder how much money I made on renewals because of the countless domains renewed in good faith but never paid. It was a gamble that I decided to take, and in some cases, it paid off, in other cases, not so much. C'est La Vie. Live and learn, all of that.

There were a few calls that were priceless. Stuff that could only come straight out of New Orleans. One of the most memorable moments during that recovery period was discovering a site with the gloriously awesome name BitchWithStyle.com (The site is no longer up, sadly).

I made this call personally. Here I am, intrepid website guy and I've gotta ask this lovely old lady if she or anyone she knew owned the domain name... BitchWithStlye.com. The woman immediately confirmed herself as the owner. This sweet lady. New Orleans never ceases to amaze and befuddle. Apparently, B.I.T.C.H. was an acronym, for:

- **B**eing
- **I**n
- **T**otal
- **C**ontrol of
- **H**erself

Apparently, she was a motivational speaker and consultant with a iconic red hat. I always liked her. She was a total sweetheart. If there's one thing you can say about these Louisiana folks, they do things their own way. You can believe that.

5. Proto point2point

Around 2008 things settled down. We'd contacted or attempted to contact everyone in every way possible short of smoke signal or hood pigeon. I figured the recovery was just about over. I was moving back to building sites and letting my staff handle the rest of the work. And how crazy is that? I had a staff now.

The staff had grown from just Peter to also Rachael who helped while her husband was in Iraq, along with our new bookkeeper Heather.

It was odd being the boss. Just a few years ago, I was this dork out of high school trying to help people get their Internet working, and now I was paying people to build websites. The team bond was cool. I'd picked a good group of weirdoes. We started to develop a corporate culture. We banded together to cook meals, and everyone began to wear the unofficial point2point uniform of camouflage pants and Hawaiian shirts. (Fashionistas, am I right?)

One afternoon after a few happy hour cocktails, we visited a martial arts supply store and ended up purchasing several bokkens, wooden training swords. Why this made sense, I can't say, but I'd suggest whiskey had a lot to do with it.

Either way, many afternoons were spent in epic sword battle in my front yard, which may or may not have attracted the attention of law enforcement on more than one occasion. Things in point2point land felt good. So many afternoons were spent moving website files, just being in the zone. We'd have loud rock and roll ripping through the speakers while folks talked, or others smacked on BBQ. Cigarettes lingered in ashtrays, and there were plenty of open beer cans. It was a punk rock business in its purest form.

But, despite all of the raucousness, as soon as the phone rang, everything would stop dead. Someone would answer the call with a very professional "thank you for calling point2point". The clients had no idea we were a total pack of misfits.

About six months after Hurricane Katrina, I started getting requests from customers that I come down for a visit. At this time, it had been a good long while since I'd been back to town. Naturally, I jumped at the chance. I wasn't a mogul or anything, just a guy running a tiny business, so I wasn't hopping on a jet to see my favorite city.

When I got to town, the New Orleans I'd tried to live in was not the New Orleans at my feet. Finding a place to stay in New Orleans post-Katrina was a lesson in frustration.

I wound up at St. Vincent's Guest House on Magazine Street. Mojo's Coffee House across the street became my New Orleans "office" where I met many of my new customers. (I occasionally see Mojo's owner, and I still thank him for allowing me to loiter as much as I did in those days. There were several evenings where I sat outside the coffee shop after closing, using Mojo's WIFI since St. Vincent's Internet was spotty at best.)

If you've never heard of St. Vincent's, I need to school you on some quick history. St. Vincent's is not your standard guest house.

St Vincent's is an old establishment. Before it's current form, it was an orphanage in 1862, and was founded by the Daughters of Charity order of nuns.

I learned this the hard way one evening when I awoke to see a small girl staring at me. Let's say I didn't get much sleep after that encounter.

When I made it down the front desk and asked about casual paranormal activities happening on the regular, I wasn't exactly met with shock. But instead, "We thought you knew? Being on the top floor, you're close to the attic. Don't worry; they're harmless".

Harmless child ghosts? Got it, no problem. New Orleans stuff, once again.

5. Laniappe

For a while there, the dust settled, and it felt like that was that. The Katrina work was over. The New Orleans clients were live, and I was running everything from Memphis. I won't say I wasn't fulfilled, or that I was wallowing in some miserable existence because I wasn't. I'd danced with the idea of moving back to New Orleans and trying to maintain some dual office thing, but really, I just didn't want to get burned. My business was a mix of the two cities, so both deserved my attention. For now, I was playing a waiting game to see how the future would unfold. One afternoon, I received a call from one of my New Orleans clients. He was looking for some pretty specific help.

"Michael, you know a lot of people, you know any good Flash designers?".

"I know an awesome Flash guy. But, he just recovered over three hundred sites from Katrina."

It turns out the gig was a contract with Department of Defense, specifically the Department of the Navy.
As a young business owner, I had specific goals. I wanted that "it" client that turned heads. I wanted a big name. I wanted a major portfolio piece.

Nothing is bigger than the military. Nothing. By now, I was interested. Landing that military gig puts you in a different league. It moves you past the local mom and pop shops and opens the doors to a lot bigger, national and international clients. Who doesn't want to play on the bigger stage? I certainly did.

The job came with a catch. I had to work in a secure facility and away from my office and away from my team. No blaring Nine Inch Nails, no office drinking. No camo and Hawaiian shirt sword fights. But, I'd get to move back to New Orleans, and this time, I'd be gainfully employed.

"You got any skeletons in your closet?"

"I got an FBI file."

"Oh boy. What did you do."

Around 2003, a hacker named Adrian Lamo was arrested by the FBI for compromising The New York Times, and effectively adding his name to the internal database of expert sources and using the paper's LexisNexis account to research some high profile people.

Adrian was known as "The Homeless Hacker" which made for an interesting manhunt until he eventually turned himself in. When Adrian turned himself in, an FBI agent took Adrian's computer and went through it with a fine toothed comb. This includes investigating every single bookmark on the favorites bar.

One of those bookmarks was to my blog. The FBI agent did a WHOIS lookup on the domain name, and my parent's address was right there. Guess who got one of those phone calls you see in the movies? I like to think he was wearing a dark suit and aviators. Since the site was registered at my parent's house, my mom answered the call. She thought it was a friend playing a joke, Mr. FBI assured her it wasn't.

I remember making a drink to steady my nerves. It was like eight thirty in the morning. I dialed the number and hoped I wasn't about to get my door kicked in all SWAT-style. The agent asks about my life, my computer experience, if I've ever met Adrian, I hadn't. Adrian was a friend of a friend. He must have appreciated my sense of humor. The agent thanks me for my time and I attempt to go about my day.

Adrian would get headlines in 2010 when he reported to the U.S. Army authorities that Chelsea Manning had claimed to have leaked a massive cache of documents, including 260,000 classified U.S diplomatic cables, to the Wikileaks organization.

I fill out the extensive background check information, send it off, and wait. They tell them I've been approved. They send over the formal agreement. I was officially heading back to New Orleans.

The problem was, I needed to figure out how to run my web hosting and domain registration business while building websites for the greatest country in the world.

I left Peter in charge of point2point. He'd earned it. He was my right-hand man. I got rid of everything. If it wasn't cheap, it was free. I bought an Amtrak ticket and packed a suitcase. Aside from my laptop, that was all I brought down. I'll never forget laying in my empty apartment the night before I got on the train. I felt so free.

But, when that train pulled into the station, it was game on. I was in New Orleans for real. No broken promises, or lost hope, at least that was the pledge I'd made to myself.

Living in New Orleans immediately after Hurricane Katrina was an eye opening experience. With so much of the city destroyed, walking the dog through the wreckage was surreal. It felt like a scene from I am Legend. Being able to see the stars at night because there were so few lights, the scary packs of dogs roaming the streets and the streets lined with refrigerators, this was a city that had taken a severe beating.

I spent the next two years working on a computer-based training program for the Navy. The job was both extremely rewarding and extremely soul crushing. The high points where meeting the seven foot tall Marines who used my software.

It was nice experiencing the praise, getting feedback that they thought it looked beautiful and also easy to use. What sucked was realizing that we were on a maintenance project, just making necessary changes and getting paid to sit there.

After those two years working for the government, I understand why our country is in the economic state it's in. The amount of contractors sitting around collecting paychecks was mind boggling. So, I ran point2point from my Navy cubicle. The work we got paid to do took me a whopping two hours a week to complete. My boss didn't care. He was busy running his DoD recruiting business from the cube next door.

I'm glad I had the opportunity to work on this project but the soul crushing cubicle lifestyle was not for me. When the contract ended, there was smoke at my heels. I was going back to Hawaiian shirts and Camo pants.

I moved to the Garden District and found the nearest coffee shop to use as a base of operations.

I didn't have a business plan. I had a business *idea*:

- Build quality sites
- Develop a good reputation
- Be affordable

It took some time, but soon enough I was working with businesses throughout New Orleans. I wasn't that kid who couldn't rub two nickels together anymore. I had an office. I was my own boss and I'd made my bones in the city of my dreams. I put my heart and soul into the city and New Orleans returned its favors.

Lessons Learned

My story isn't all doom and gloom. It's pretty positive. It's a Cinderella story, a silver lining on an angry gray cloud named Katrina.

Thanks to the experience, though, I'm passionate about preventative maintenance along with the pre-emptive checking of systems.

Make sure you can answer these questions because if you can't you're in for a lot of work trying to get back online should something go wrong.

Step 1: Who's my domain registrar, who's my administrative contact, and when do I need to renew?

Many people are unsure where their domain name is registered. Many times I've found that business owners set up their domain registration when starting their organization and are only reminded of it when the domain expires.

Do what's called a WHOIS lookup which will tell you where your domain is registered and also will show you who technically owns your domain name.

Many web hosts, including the one that I pulled from the waters of Katrina, register the domain name in their name instead of the customers. This doesn't cause a problem until their servers are under seven feet of water and you can't access your domain to point it to a server that isn't soaked.

A WHOIS lookup can be performed at whois.icann.org/en and click WHOIS Lookup under Tools. This will tell you where the domain is registered, who's listed as the Admin, where the website is hosted and when it'll expire.

Once a domain name expires the Web site will display a "coming soon" or "wish to purchase" page, and all email connectivity will cease. Knowing when your domain expires is crucial to avoid primary business connectivity.

Step 2: Does my web host have survivability?

This is nearly impossible for a client to figure out until it's too late. Every web host will assure you that they're making daily backups, but it's difficult to test your web host without asking them to restore the website. It might cost you, but it's worth it once a year to ask your host to restore to a backup - just to confirm they can.

Step 3: Local backups of email and website

To guarantee survivability, make sure nothing is stored in a 3rd party. You can't trust anyone with your data except yourself.

Ask your web host how to download mission critical data to your network and store it on jump drives in safety deposit boxes. It's the final solution to situations like Hurricane Katrina, and if this data is important, then it needs to be secured.

Hurricane Katrina was the wake-up call that New Orleans needed regarding survivability. People assumed things would be okay because "so and so" was handling it when they weren't. Test your backups. Once you tested them, test them again, and then one more time.

Having your domain name registered with your IT company and not yourself is a disaster waiting to happen. If the IT company goes out of business, you've lost intellectual property that can never be recovered.

You can't go above them and contact the company they registered your domain with because it doesn't matter since they still own your domain name.

It's your responsibility to know if you own your domain name or if your IT company which is one hurricane away from being gone, owns the irreplaceable object which you is your domain name.

Have further questions? Let's talk. I'll even bring you a Hawaiian shirt.

Web Presence Check List

1. My Domain Registrar is _____

    a. It next expires on _____

    b. My contact information is listed (Y/N)

2. My Web Site Host is _____

    a. My DNS is managed by_____

3. My Email Host is _____

    a. My Email Host is managed by

    _____